VOLUME 3

The LEGO® Technic Idea Book

FANTASTIC CONTRAPTIONS

Yoshihito Isogawa

no starch press

The LEGO® Technic Idea Book: Fantastic Contraptions. Copyright © 2011 by Yoshihito Isogawa.

Printed in Canada

15 14 13 12 11 10   1 2 3 4 5 6 7 8 9

Mixed Sources

Cert no. SW-COC-001271
© 1996 FSC

FSC

ISBN-10: 1-59327-279-0
ISBN-13: 978-1-59327-279-1

Publisher: William Pollock
Production Editor: Serena Yang
Cover and Interior Design: Yoshihito Isogawa and Octopod Studios
Technical Reviewer: Sumiko Hirano
Compositor: Octopod Studios

For information on book distributors or translations, please contact No Starch Press, Inc. directly:
No Starch Press, Inc.
38 Ringold Street, San Francisco, CA 94103
phone: 415.863.9900; fax: 415.863.9950; info@nostarch.com; www.nostarch.com

Library of Congress Cataloging-in-Publication Data

Isogawa, Yoshihito.
 Fantastic contraptions / by Yoshihito Isogawa.
     p. cm. -- (The LEGO Technic idea book ; 3)
 ISBN-13: 978-1-59327-279-1
 ISBN-10: 1-59327-279-0
 1. Mobile robots. 2. LEGO toys. I. Title.
 TJ211.415.I86 2010
 629.8'932--dc22
                              2010031342

This book is full of little seeds for ideas.

It is you who will cultivate those seeds

so they grow into wonderful masterpieces.

Yoshihito Isogawa

LEGO® Technic is designed to allow builders to create advanced models with moving parts, like those built with LEGO MINDSTORMS®. The *LEGO Technic Idea Book* series is a collection of unofficial LEGO building guides that offer hundreds of ideas and examples for building mechanisms with Technic. This volume focuses on using propellers, weights, magnets, and other parts to create all sorts of moving contraptions.

Building with LEGO

LEGO bricks aren't designed to fit in just one specific place, one particular way. Your imagination is your guide when building with LEGO, and you can put bricks and other LEGO pieces together in many ways to build an almost infinite number of creations. After building a model according to the instructions included with your set, try taking the model apart and using its pieces to create a variation of the model—or build a new one altogether. That's where the real world of LEGO begins.

My hope is that this book will give you some ideas to help you build your own original creations.

You Are the Creator

The LEGO Technic Idea Book: Fantastic Contraptions is full of photographs of mini-projects designed to show you various ways to build with LEGO bricks. Combine these projects, add decorations, and change them to create your own unique masterpieces.

The Use of Color

The examples in this book are made with parts of various colors to make it easier for you to see the individual brick shapes. Judicious use of color can add real beauty to your models, and I've tried, wherever possible, to use colors in an artistic way. You don't need to use the colors I've chosen in your models; use whichever colors you want to use to make your projects your own.

Where Are the Words?

Other than this brief introduction and the table of contents, this book has almost no words. Instead, you'll find a series of photos of increasingly complex models that are designed to demonstrate building techniques. This is an idea book; it's about imagination. Rather than tell you what to see or think when you look at each photograph, I encourage you to interpret each one in your own way. If I were to tell you what to see, you would see things through my eyes. My hope is that you will see my models through your own eyes and that your interpretations will lead you to invent your own LEGO creations.

JOIN THE DISCUSSION!

View videos of many of this book's models, ask questions, and share your own designs at **http://nostarch.com/technic/**.

Praise Your Child

When your child shows you their creations, take the time to really look at them together. Ask your child what they were focused on when building their model or what they wanted to accomplish. Offer your child sincere praise about their work and address aspects of the model that impress or surprise you. Talent is fostered by praise. Encourage and praise your child, and watch their talent shine through.

Express Your Feelings

Talk to your child about their creations. Ask them to show you how things move and how the parts fit together. Have them explain how they came up with their design. Your words can serve as hints or advice for your child, planting the seeds for new ideas.

Play with Your Child

Make things with your child. Offer ideas and build together. For a challenge, compete against each other to build different versions of a model. It can be inspiring for your child to see what an adult can do. When competing with your child, always encourage them and explain your creations so that they can learn from your experience.

THE *LEGO TECHNIC IDEA BOOK* SERIES

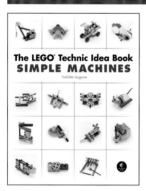

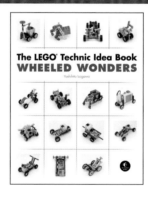

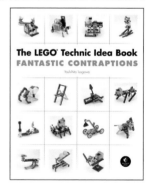

Contents

Part 1

Part 2

Part 4

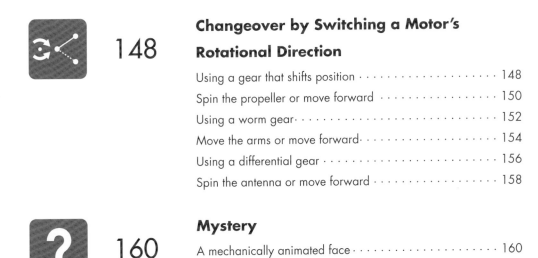

PART 1

 4

 12

2 16

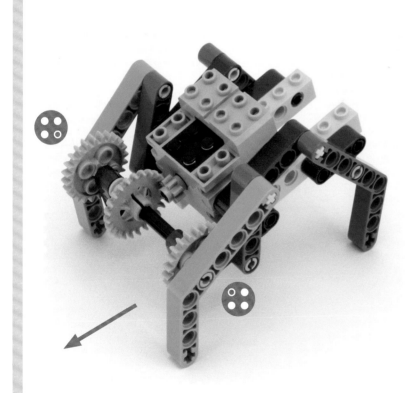

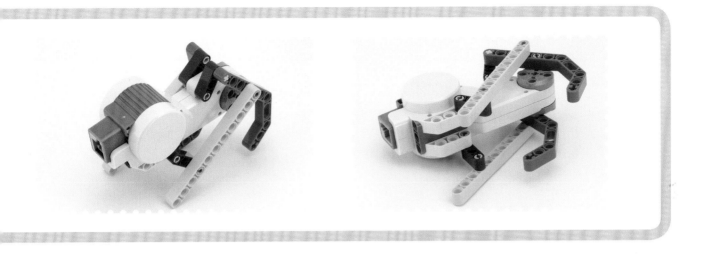

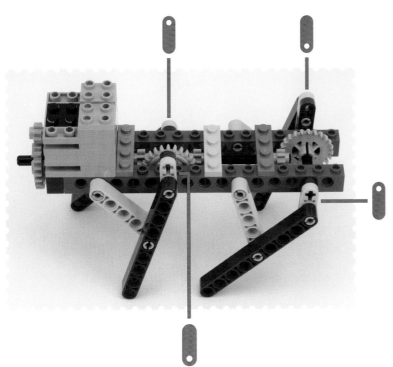

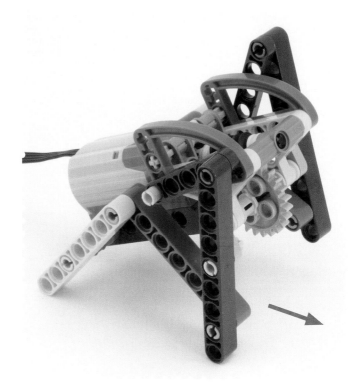

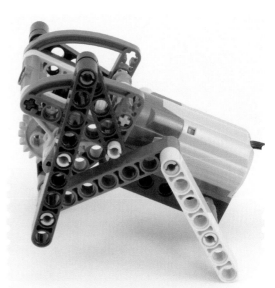

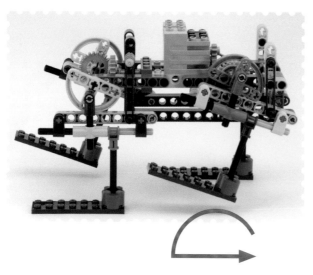

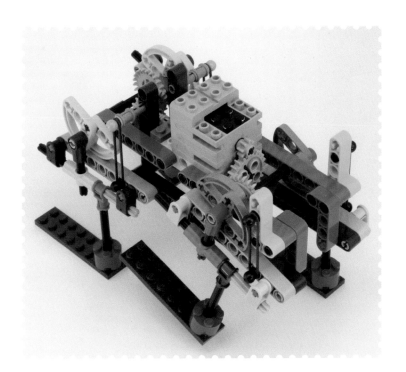

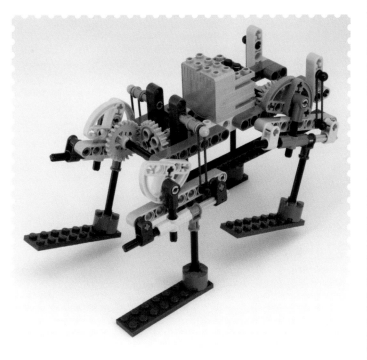

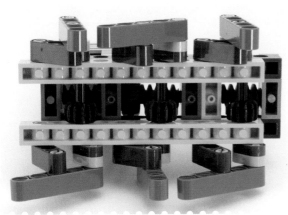

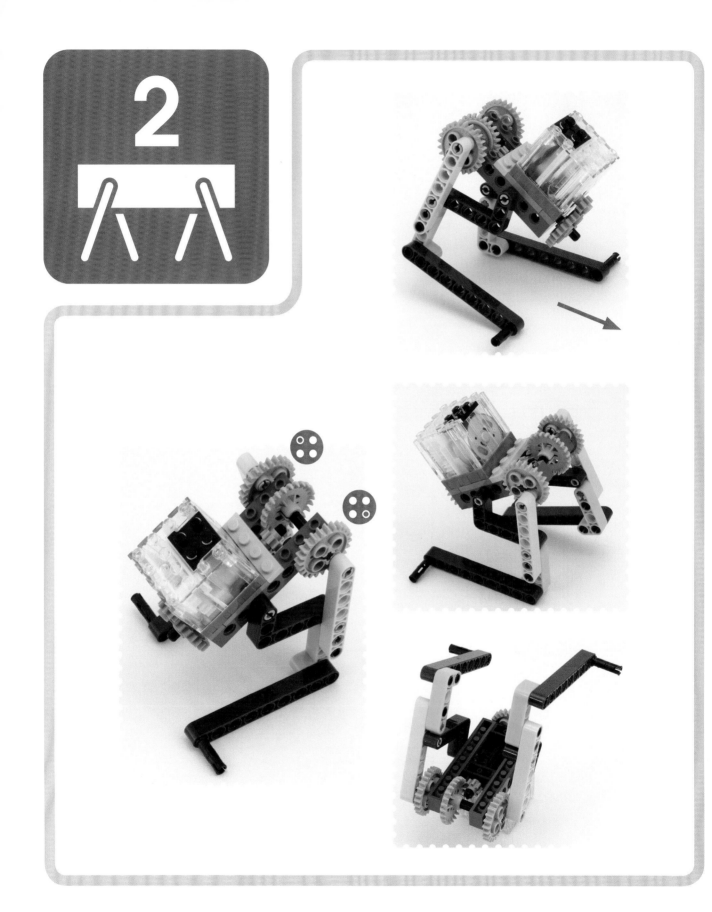

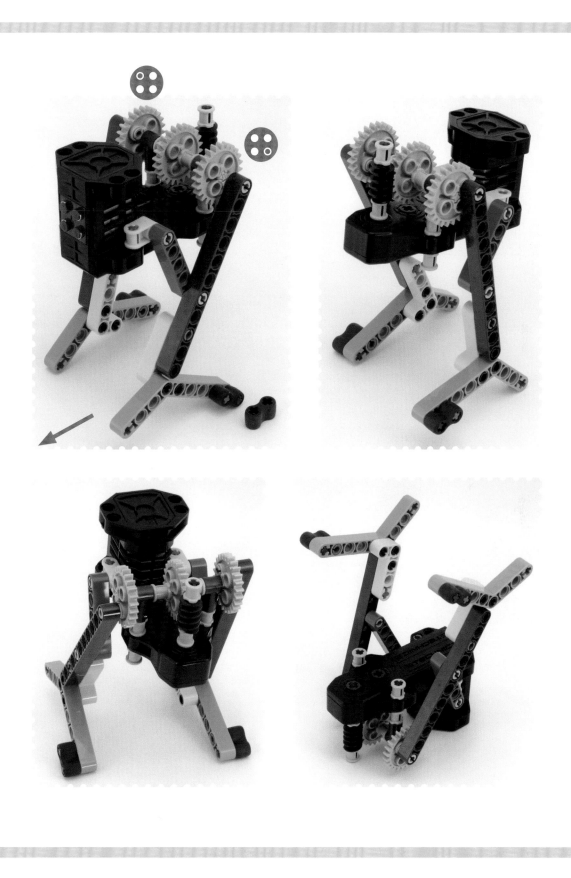

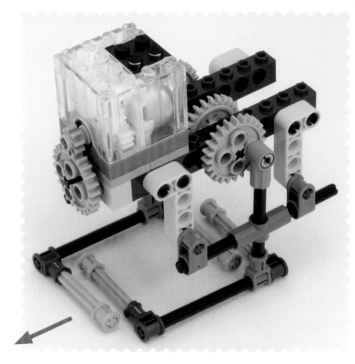

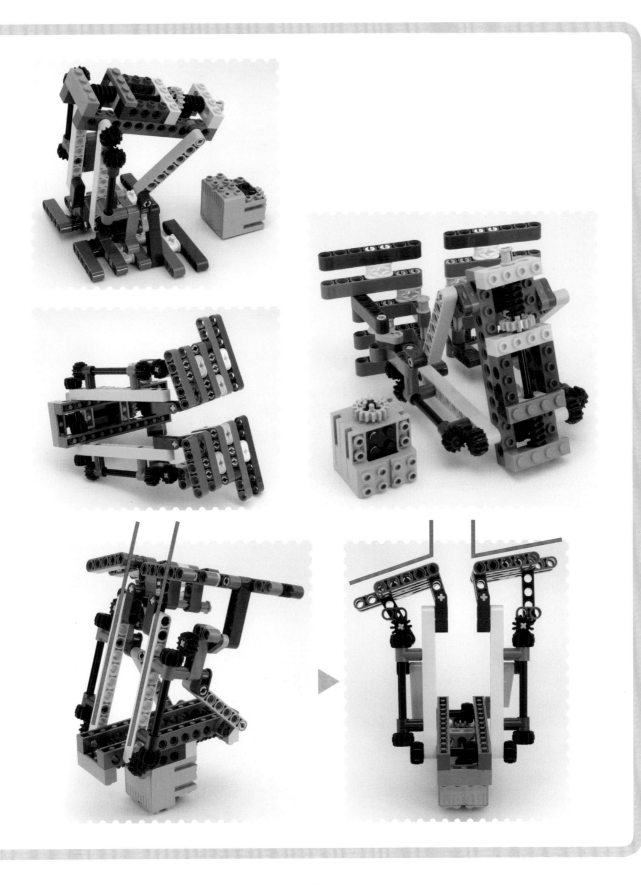

PART 2

 24

38

46

 56

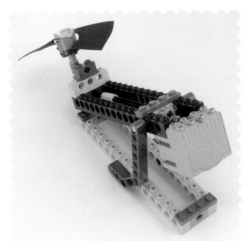

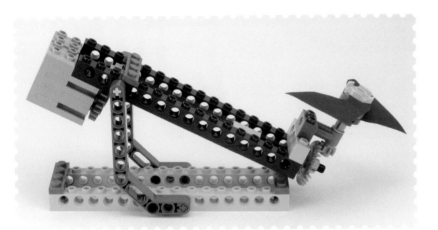

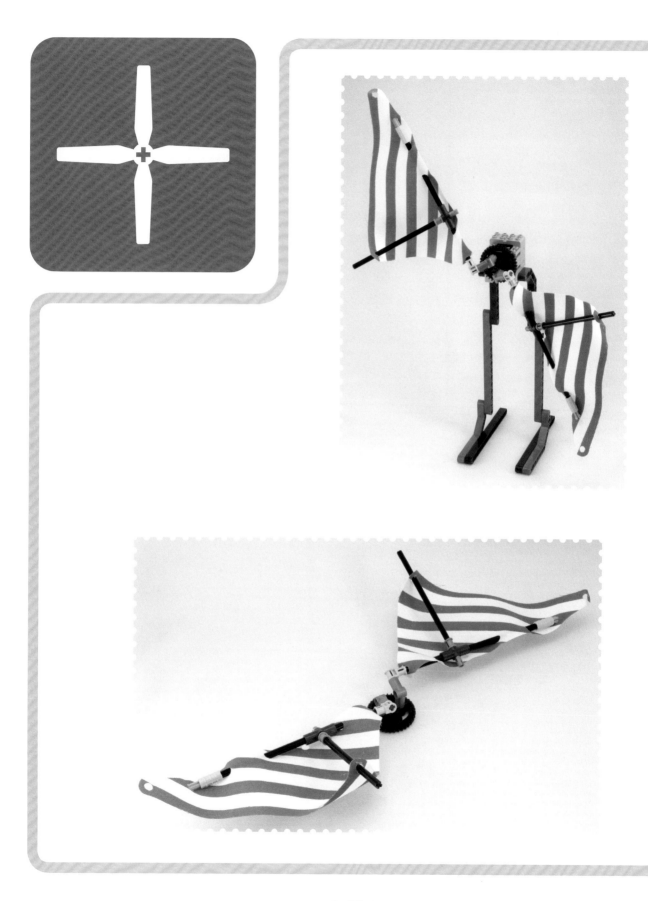

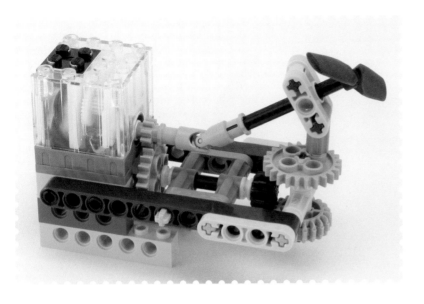

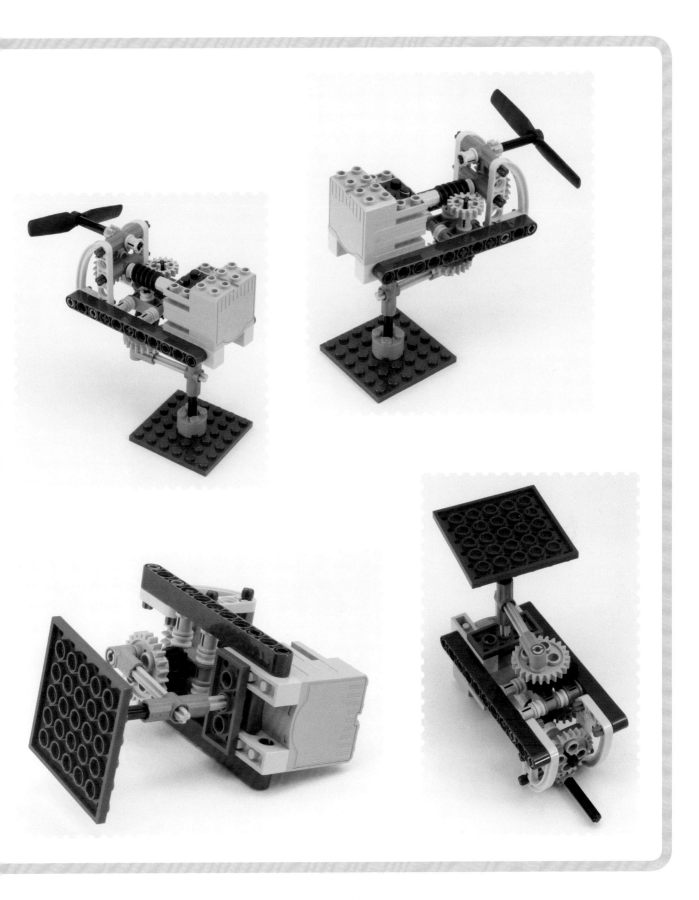

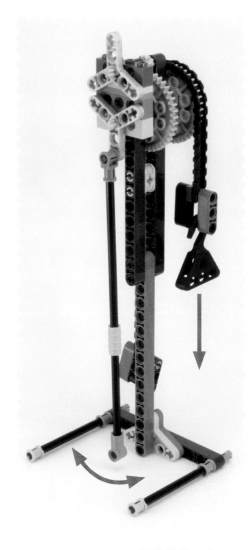

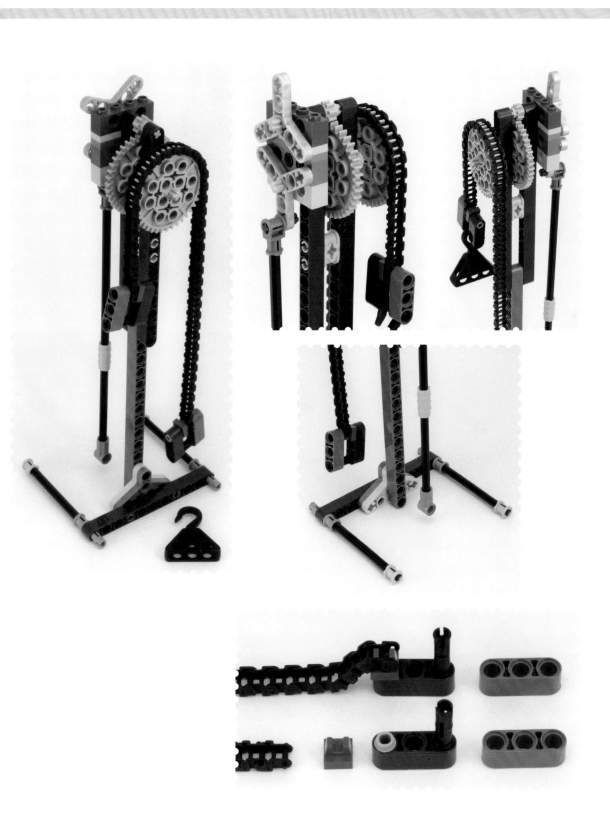

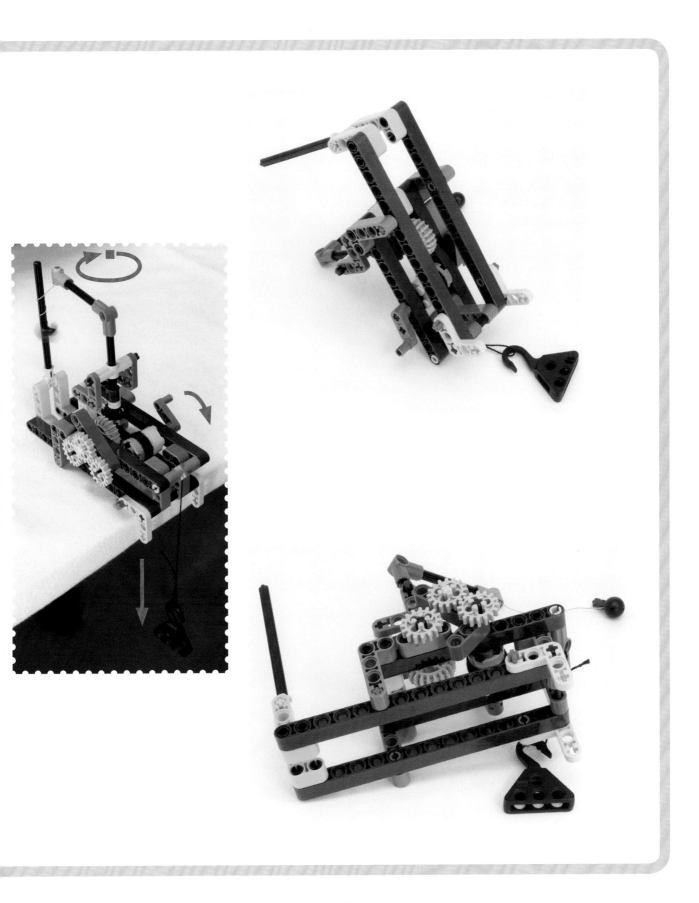

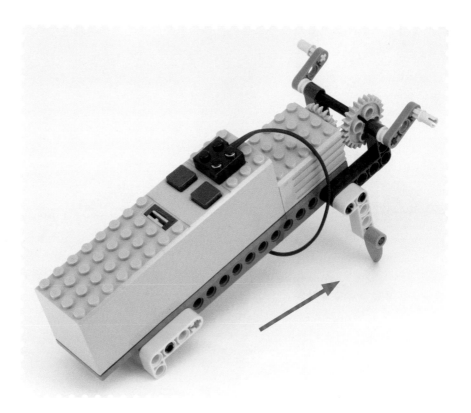

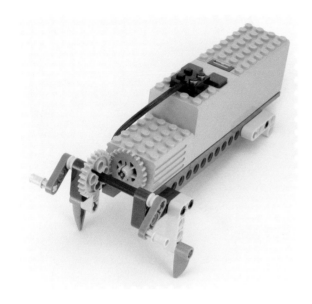

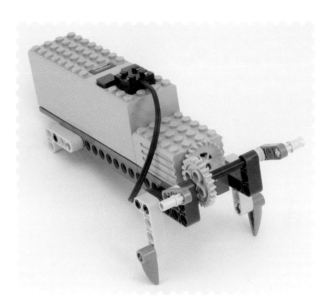

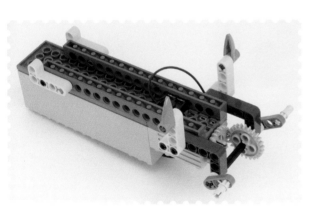

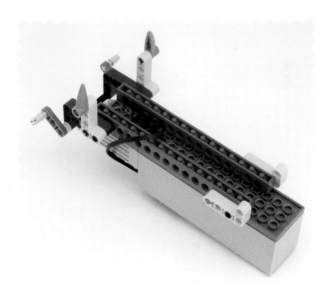

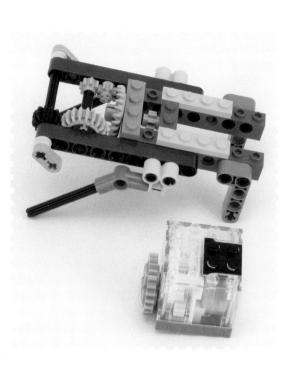

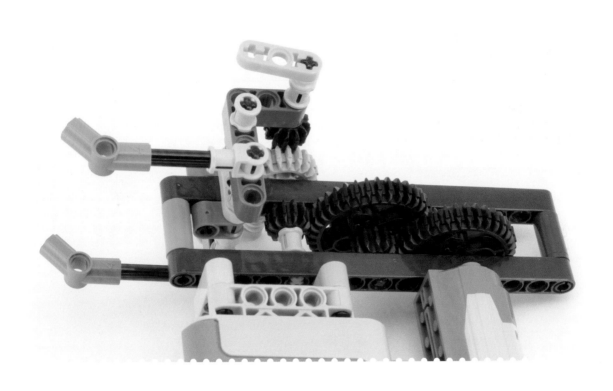

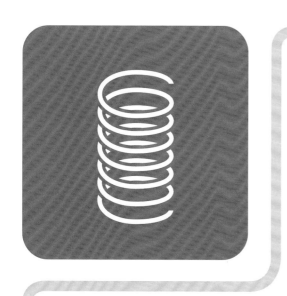

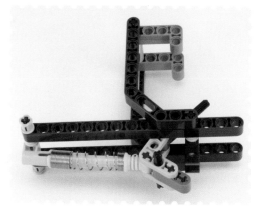

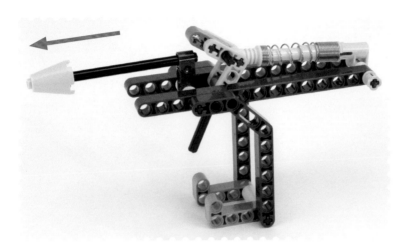

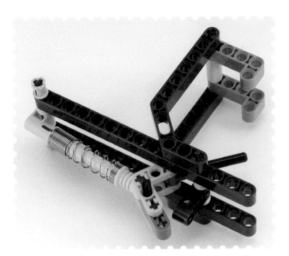

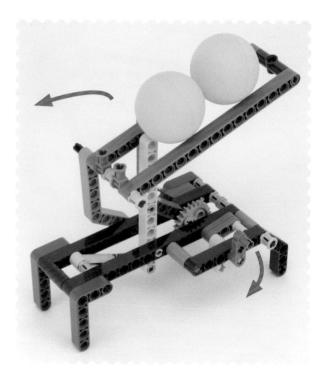

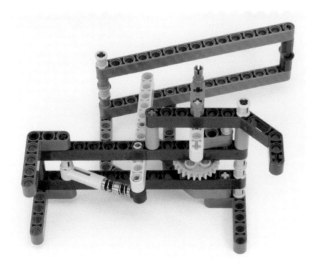

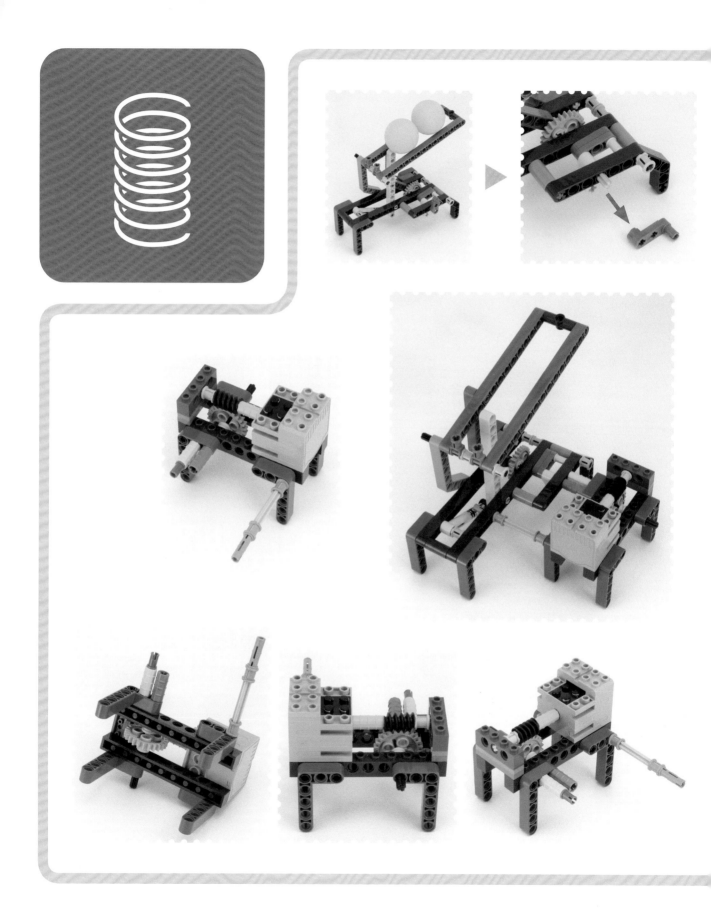

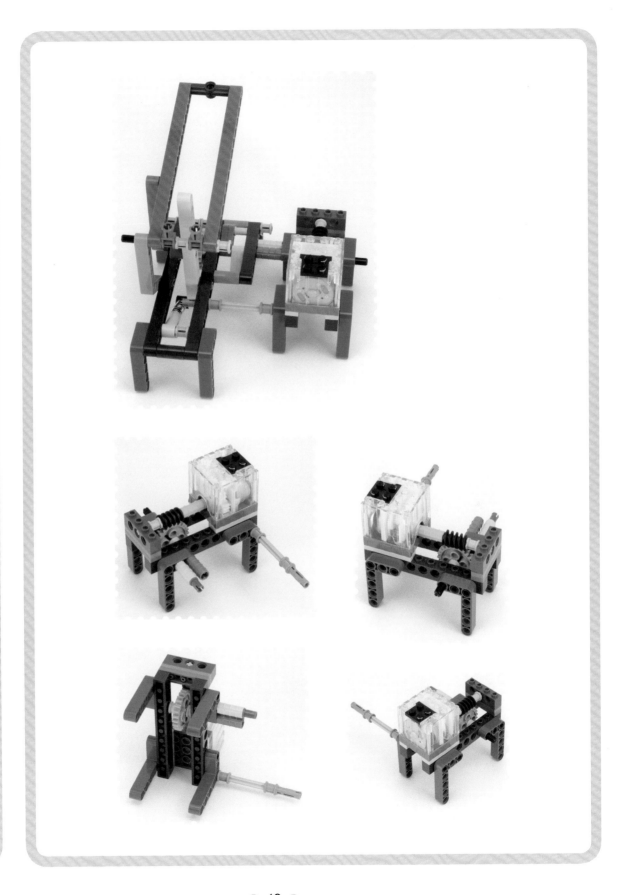

PART 3

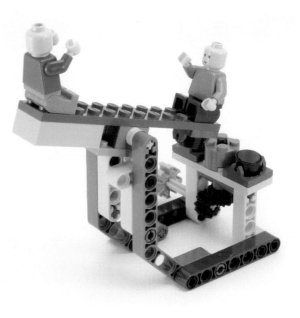

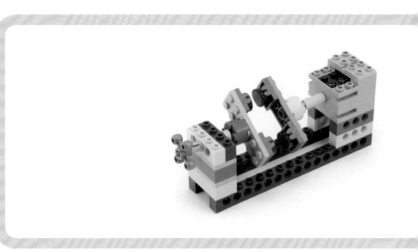

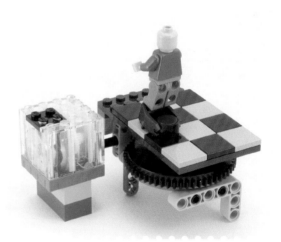

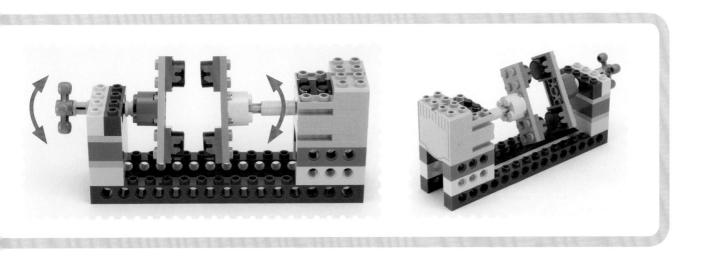

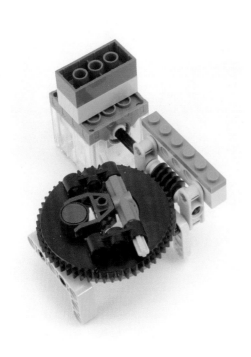

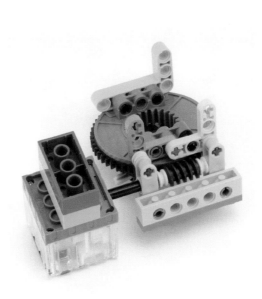

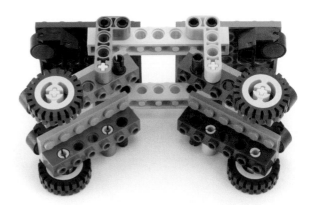

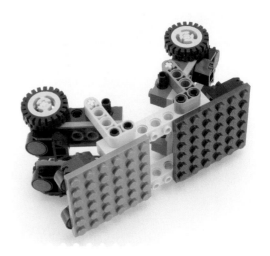

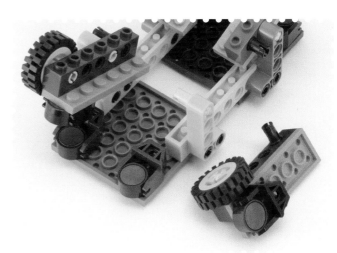

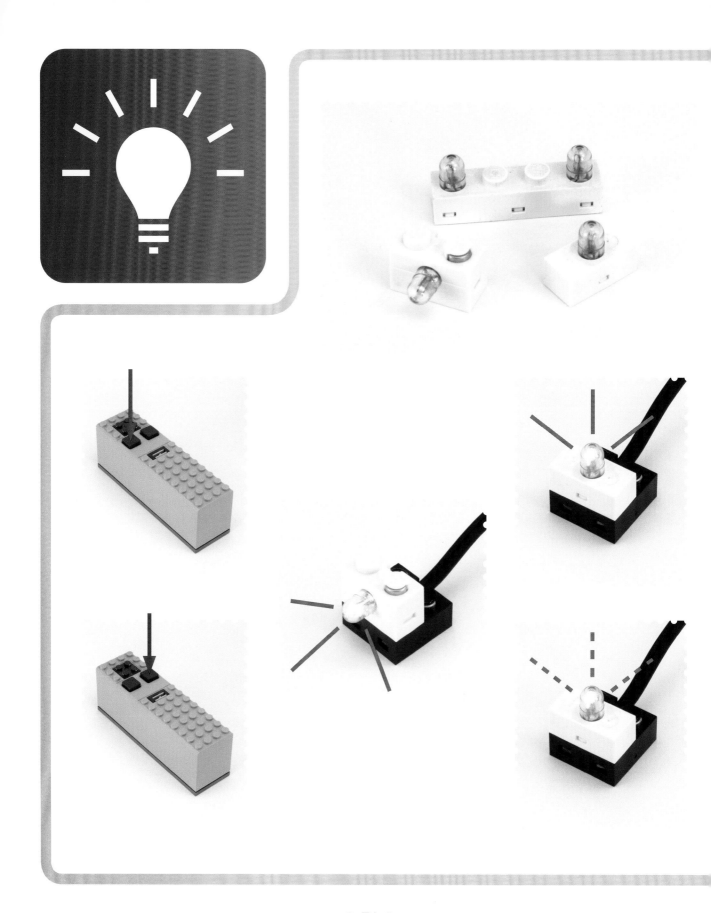

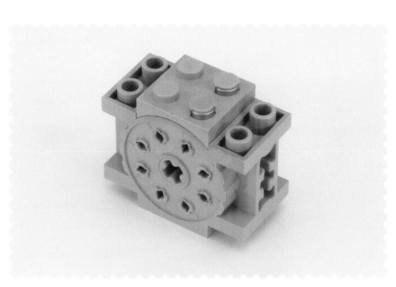

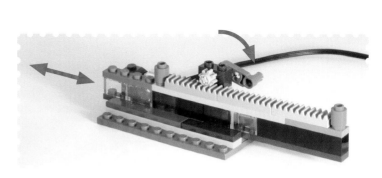

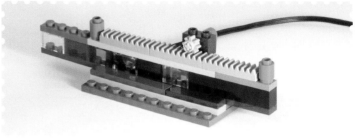

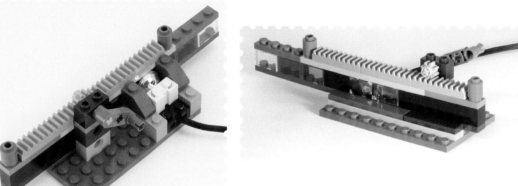

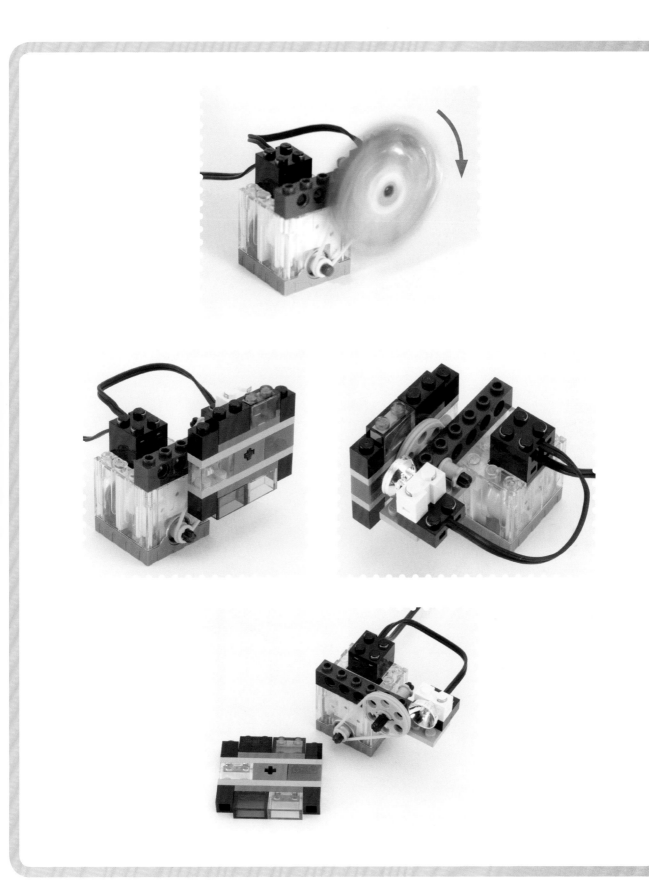

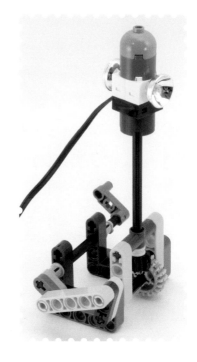

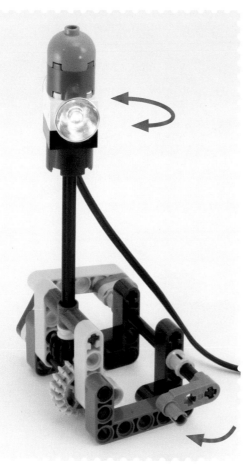

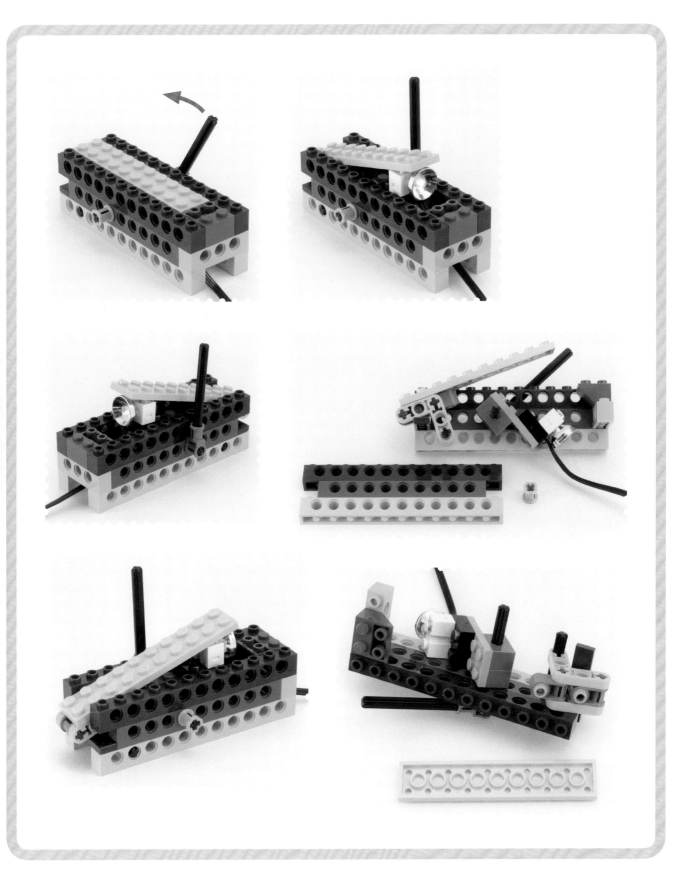

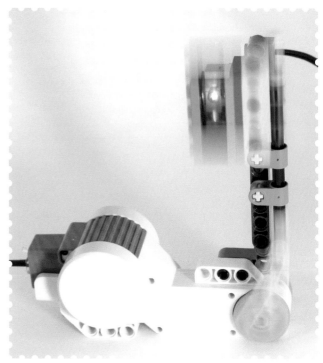

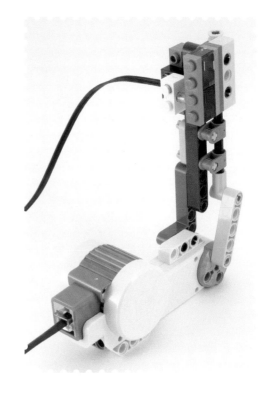

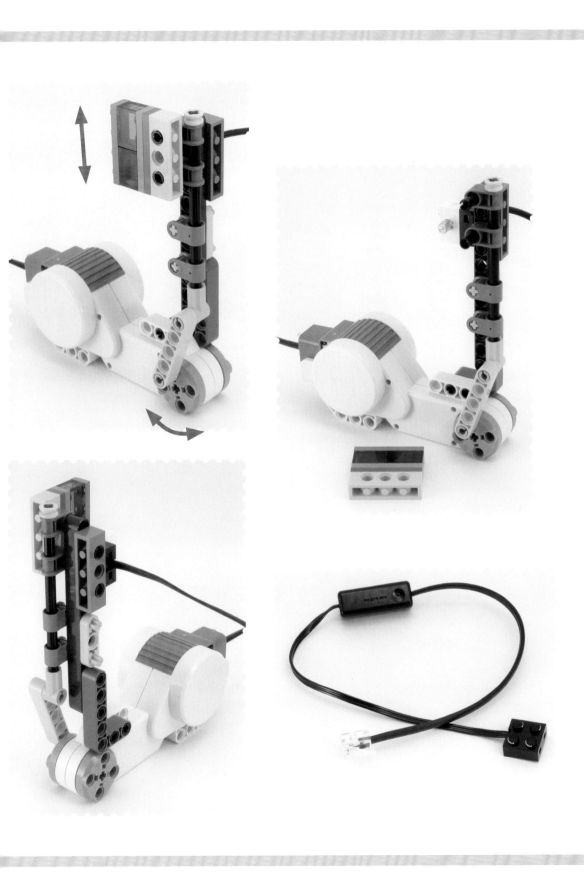

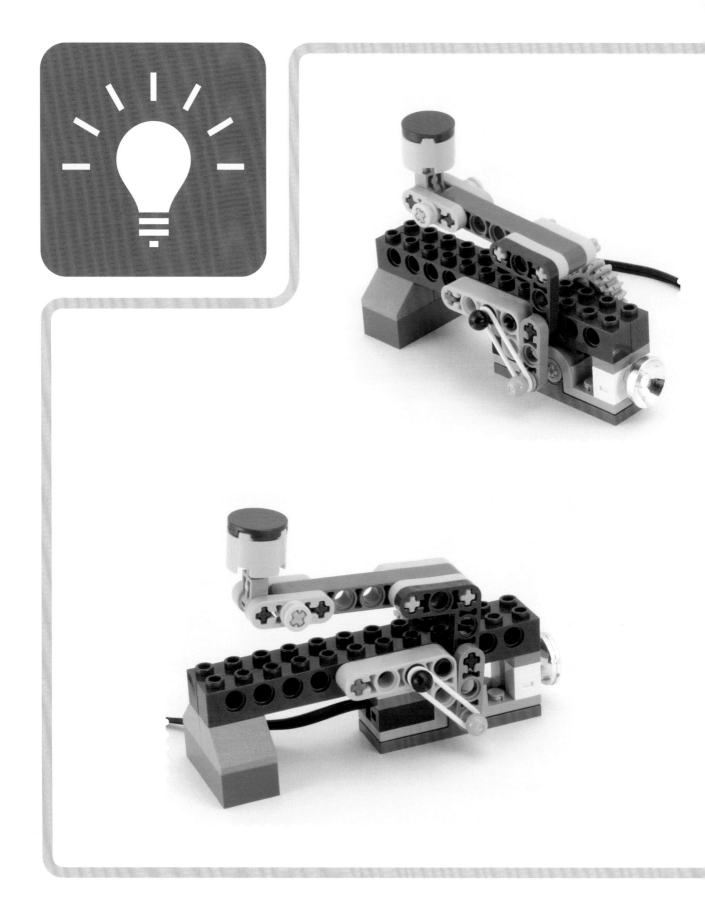

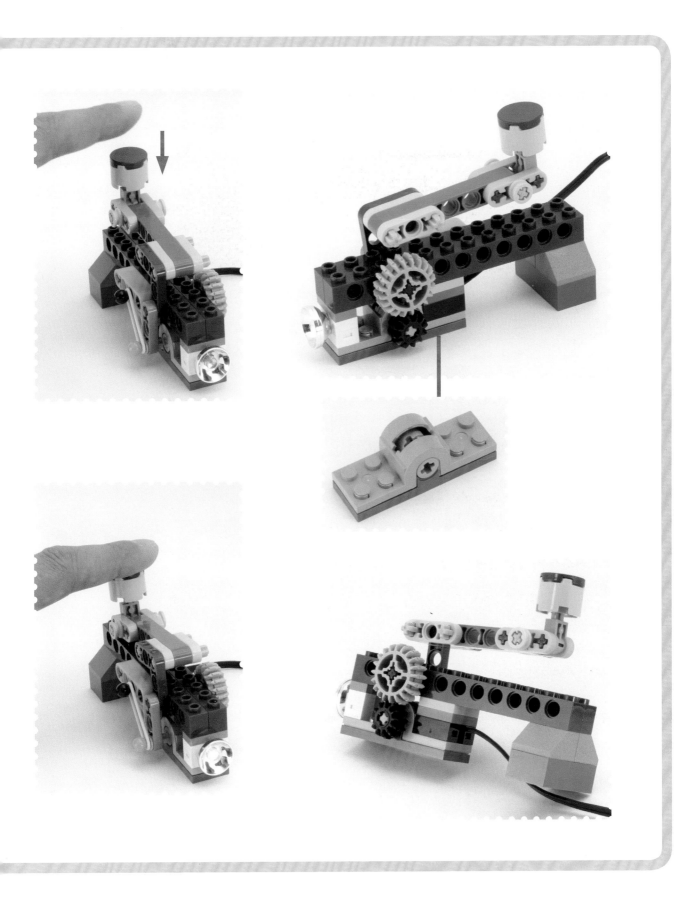

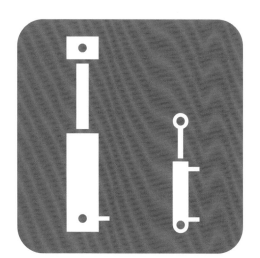

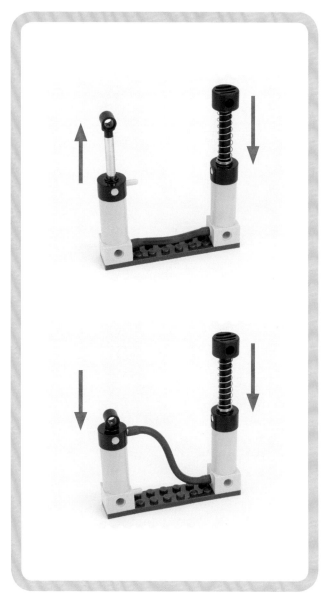

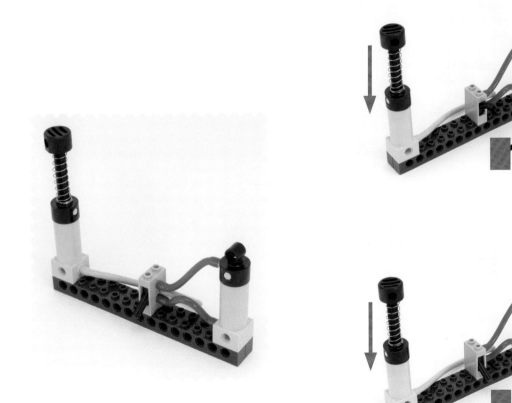

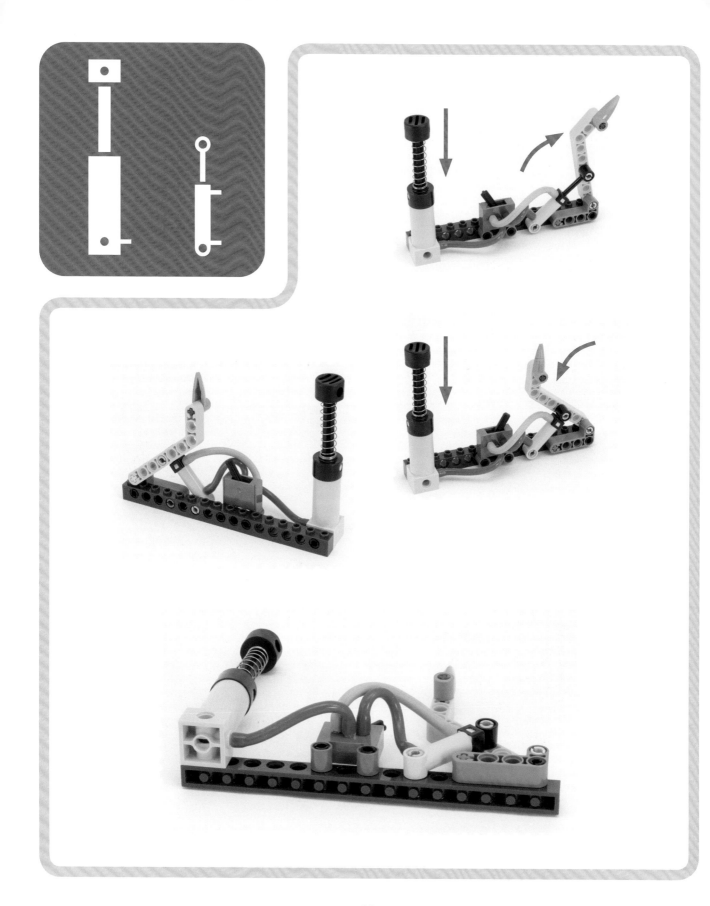

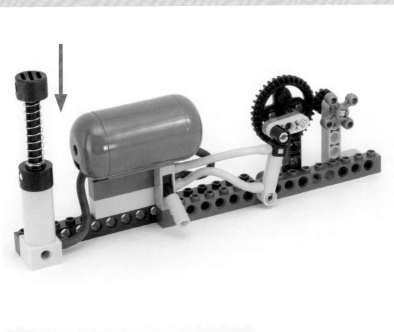

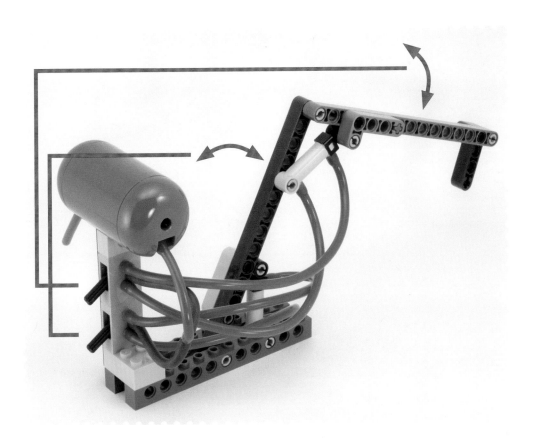

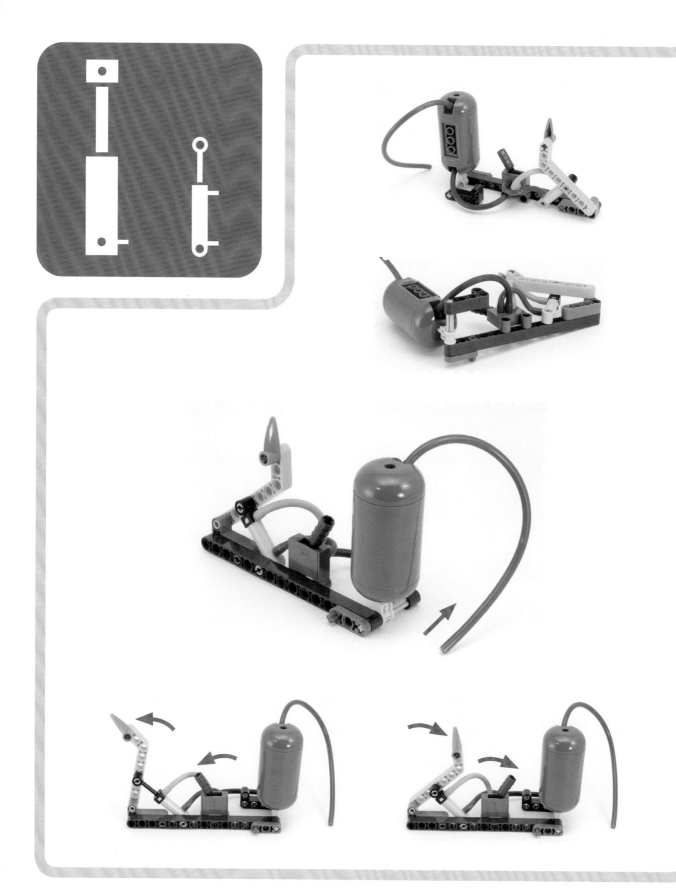

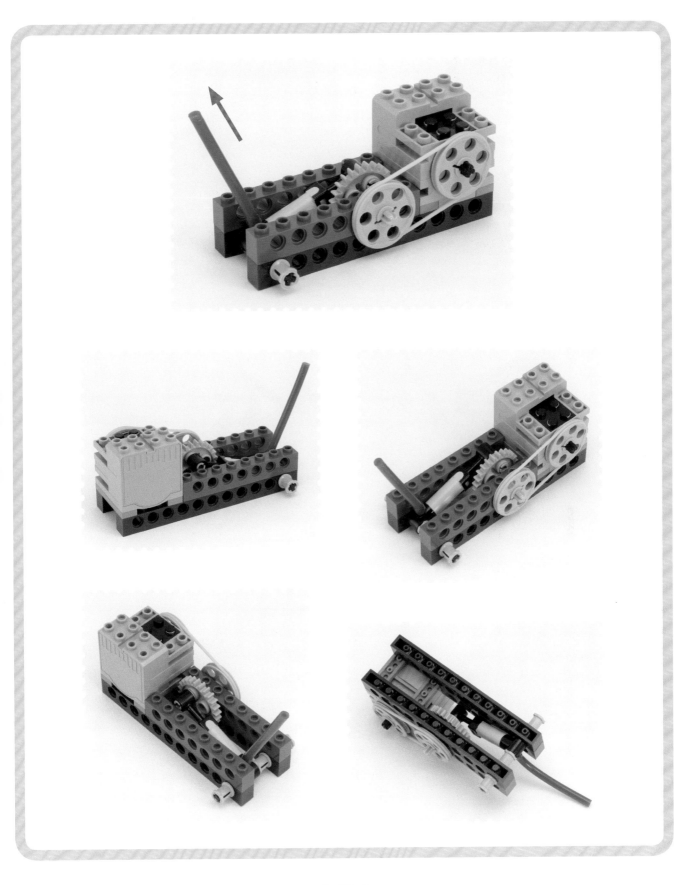

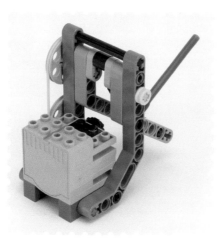

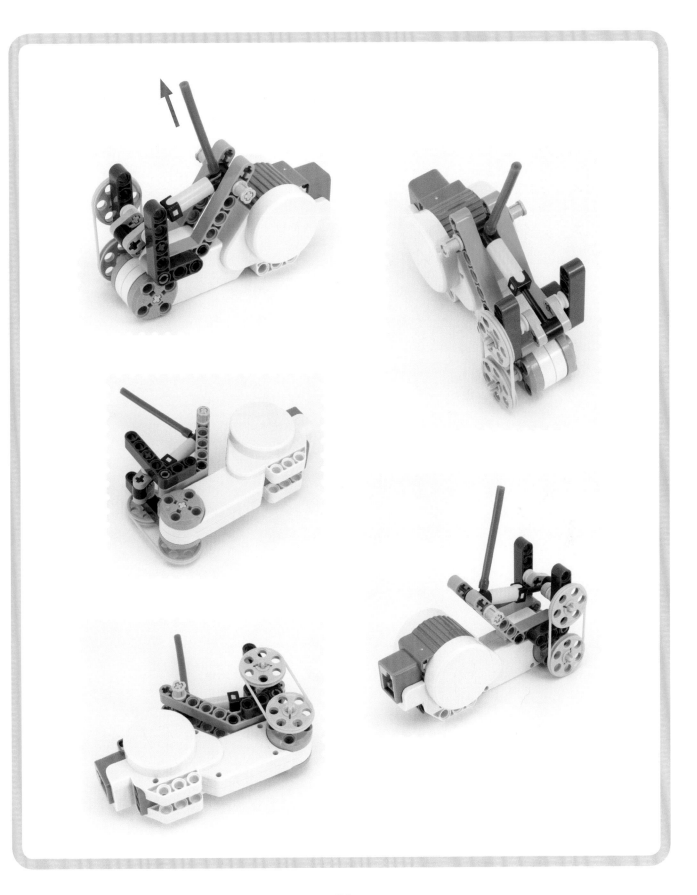

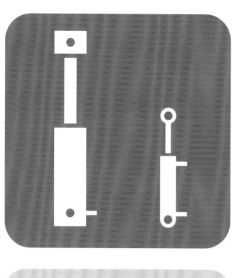

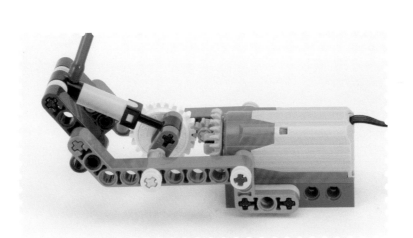

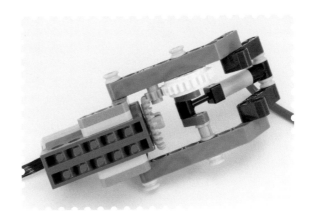

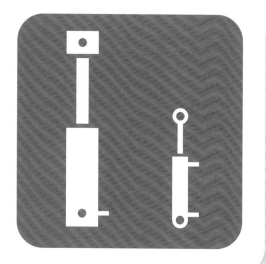

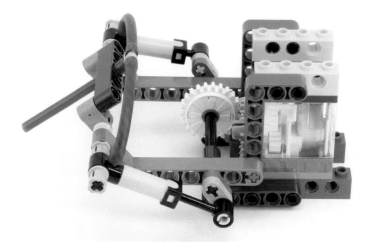

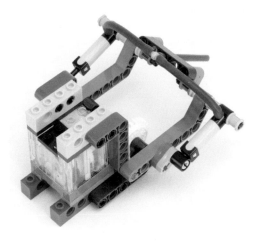

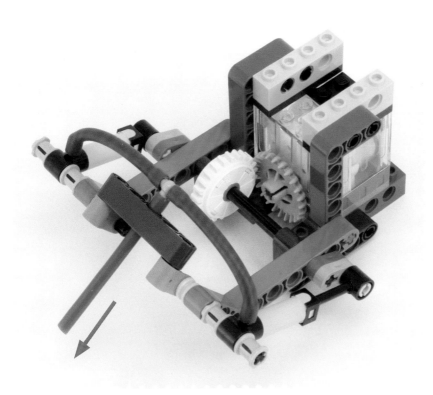

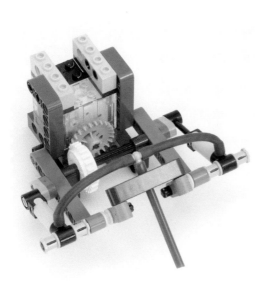

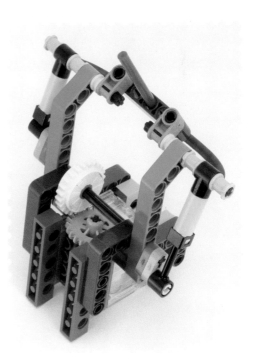

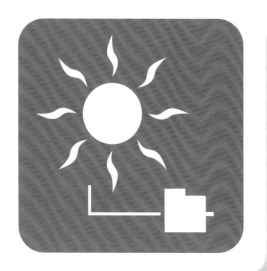

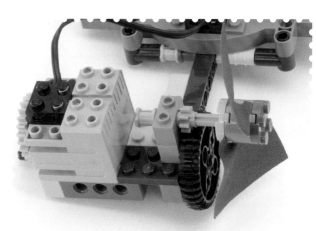

PART 4

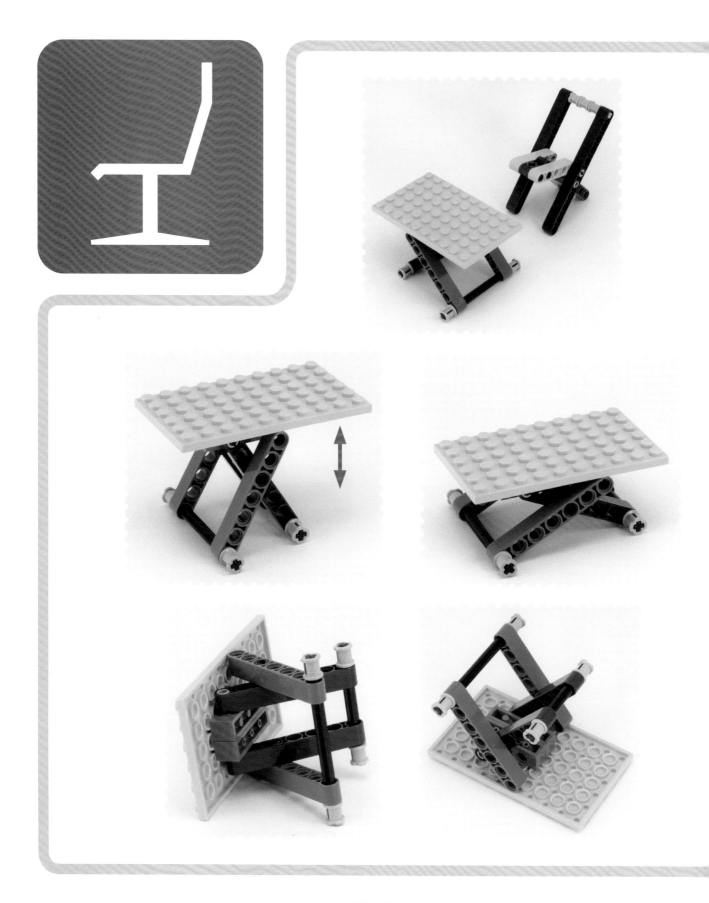

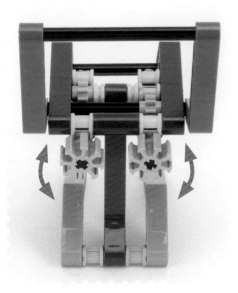

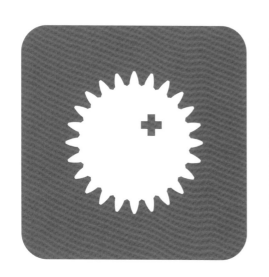

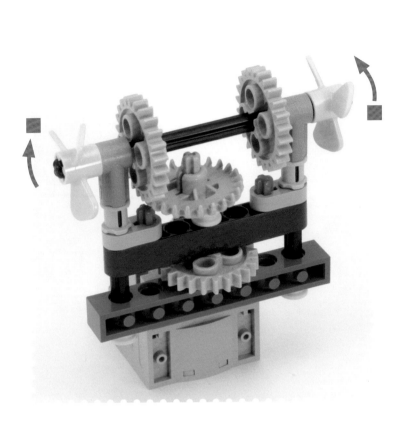

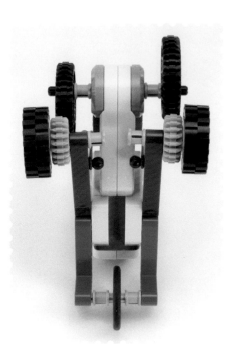

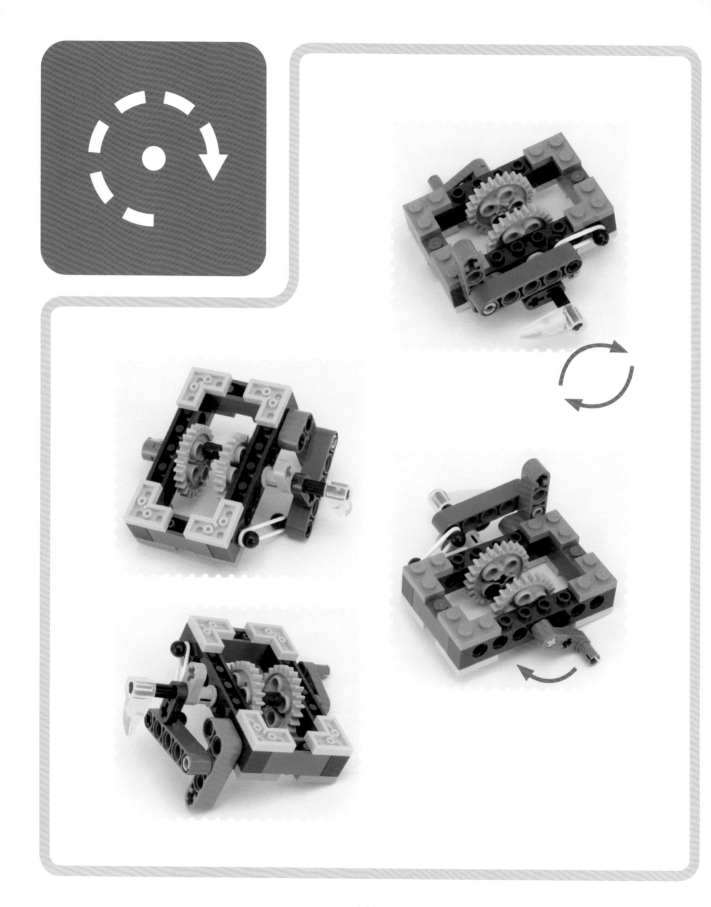

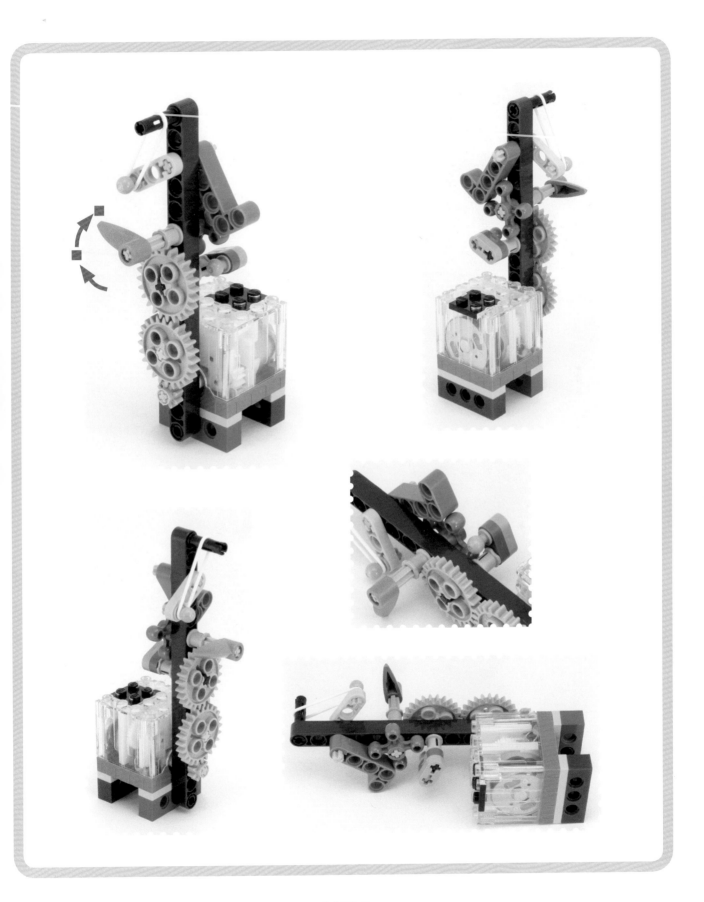

134

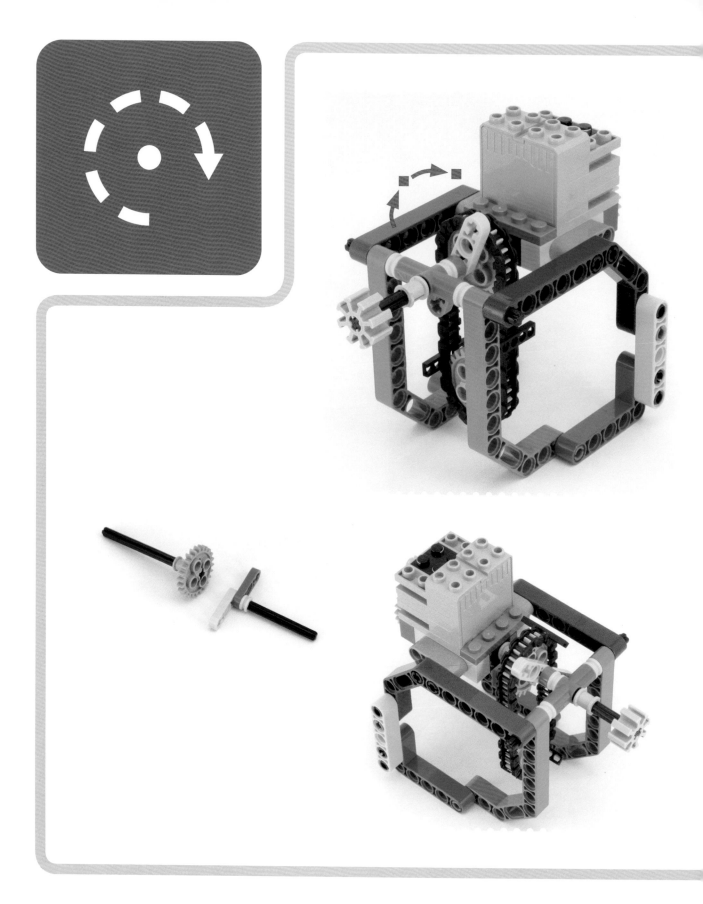

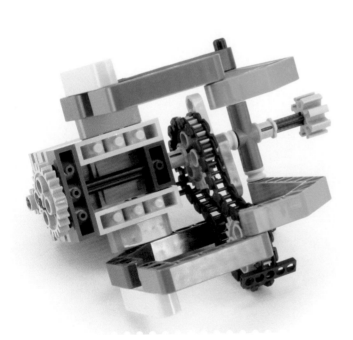

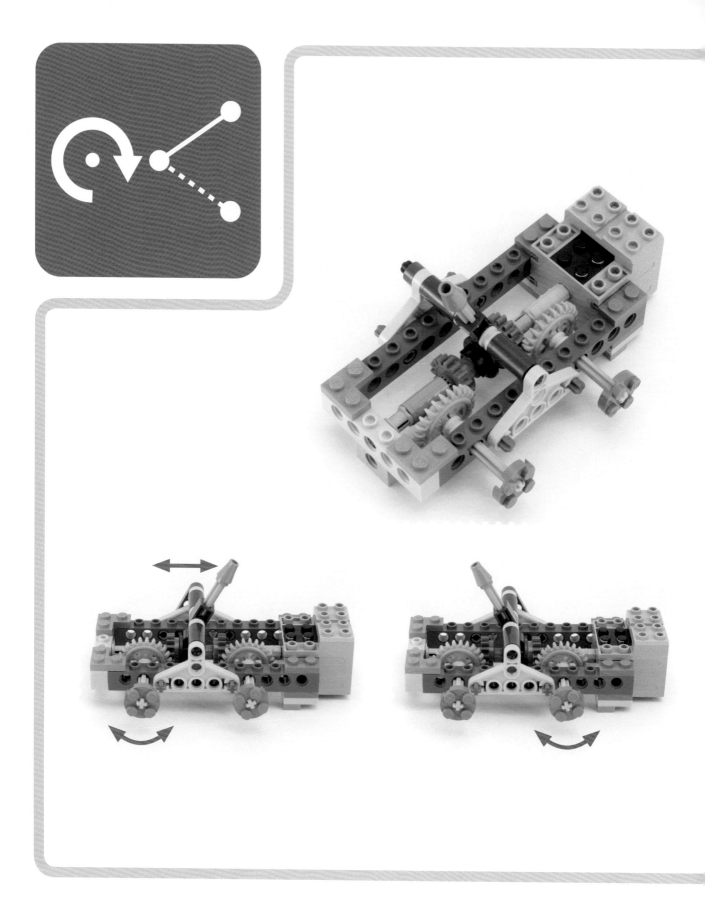

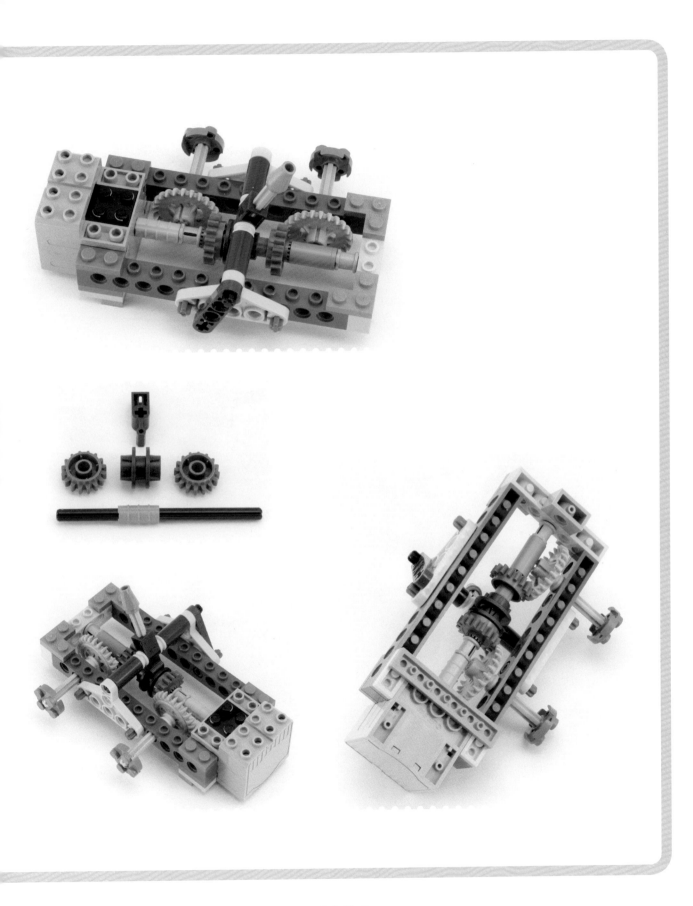

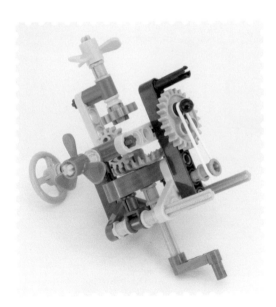

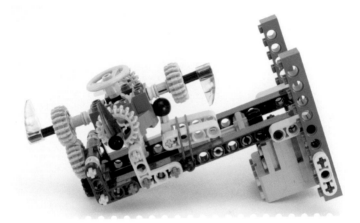

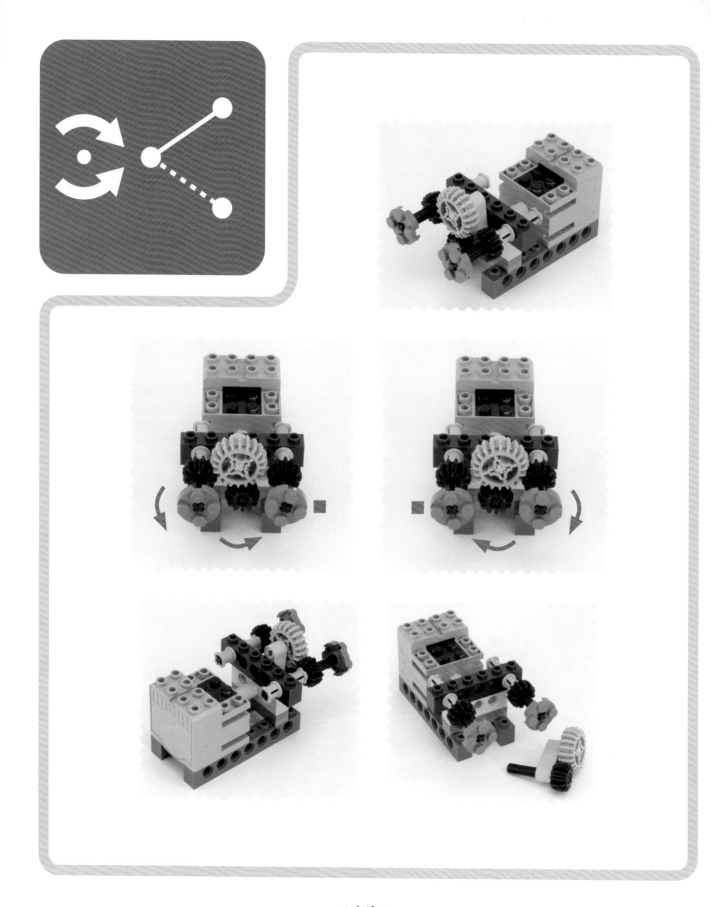

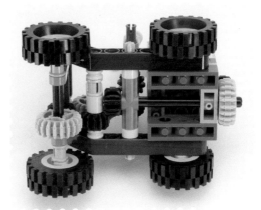

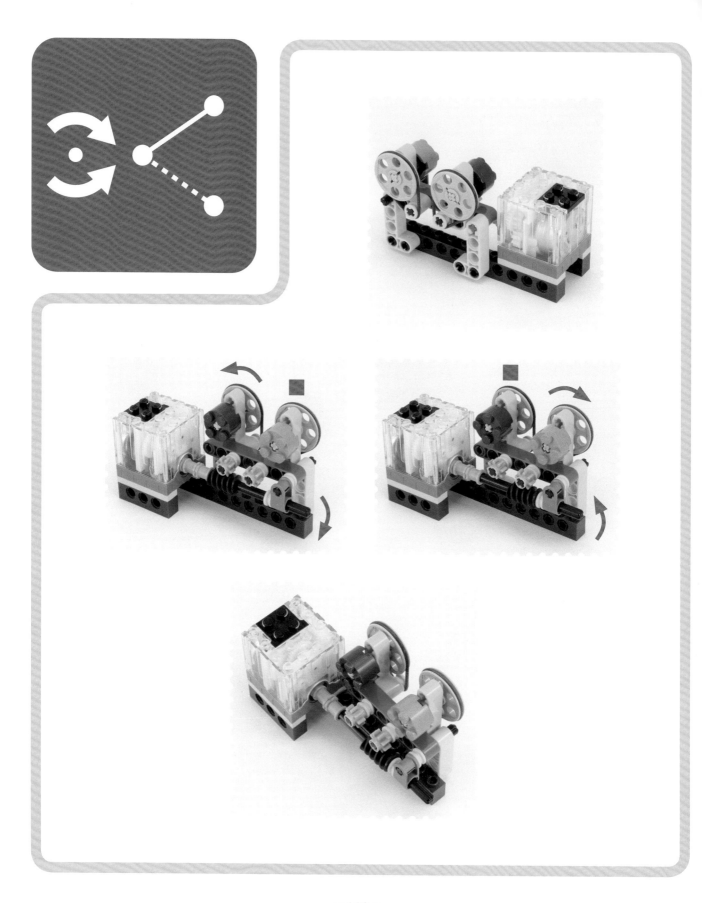

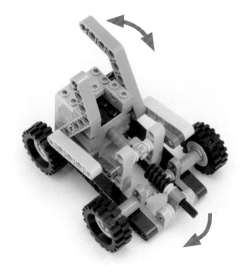

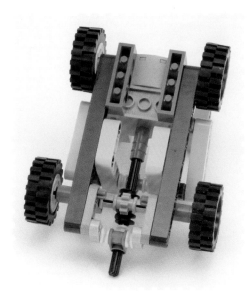

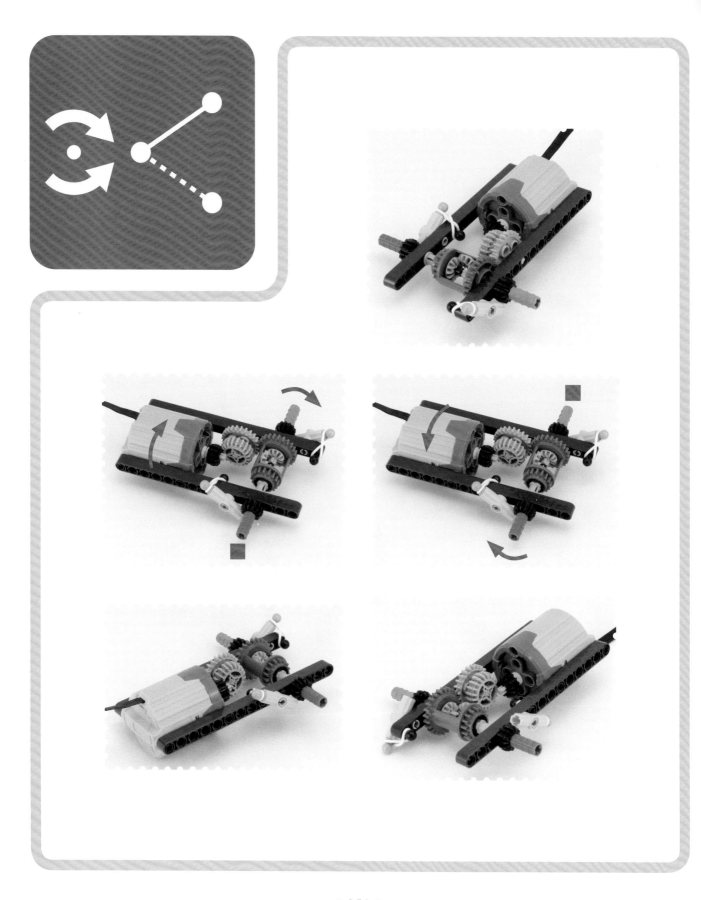

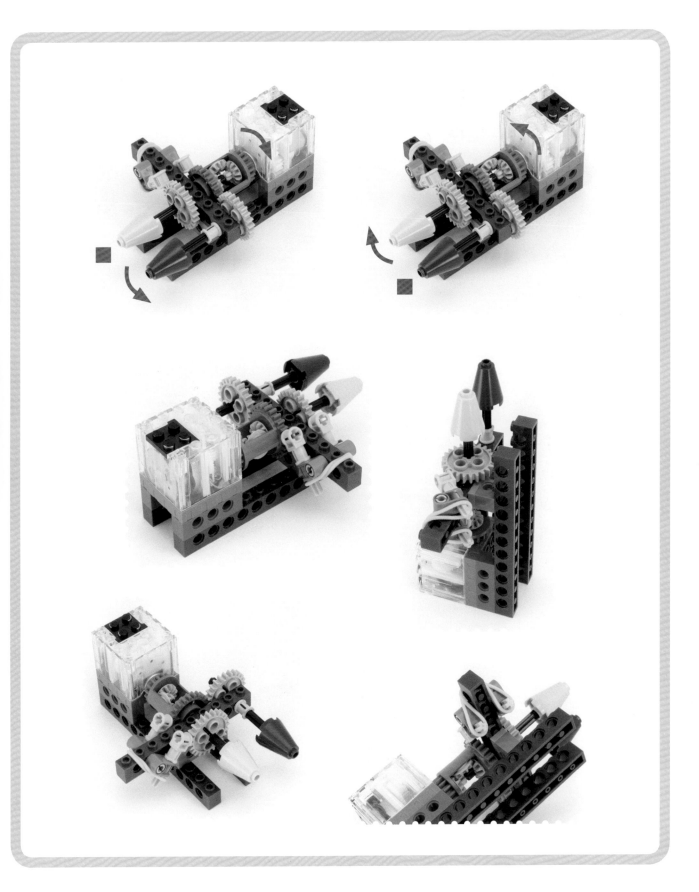

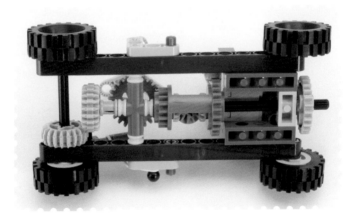

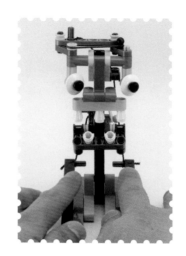

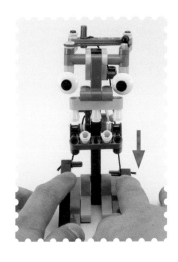

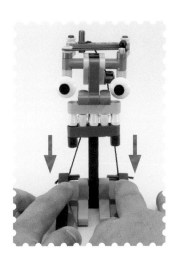

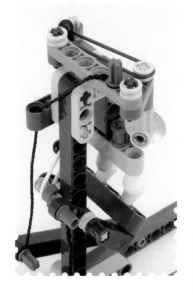

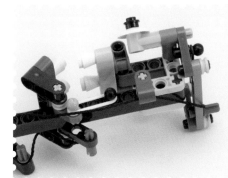

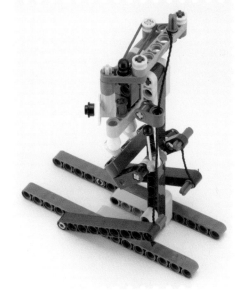